EXPLORING THE UNIVERSE

Stars and Galaxies

ROBIN KERROD

**RAINTREE
STECK-VAUGHN
PUBLISHERS**

A Harcourt Company

Austin   New York
www.raintreesteckvaughn.com

**First American edition published in 2002
by Raintree Steck-Vaughn Publishers**

© 2002 by Graham Beehag Books

All U.S. rights reserved. No part of this book may be reproduced, stored in a retrieval system, or transmitted in any form or by any means—electronic, mechanical, photocopying, recording, or otherwise—without the prior written permission of Raintree Steck-Vaughn Publishers, except for the inclusion of brief quotations in an acknowledged review.

Raintree Steck-Vaughn Publishers
4515 Seton Center Parkway
Austin, Texas 78755

Website address: www.raintreesteckvaughn.com

Library of Congress Cataloging-in-Publication Data is available upon request.

ISBN: 0-7398-2816-9

Printed and bound in the United States.

1 2 3 4 5 6 7 8 9 0 05 04 03 02 01

Contents

Introduction 4

Distant Suns 6

Matters of Life and Death 18

Galaxies of Stars 30

The Mighty Universe 38

Glossary 44

Important Dates 46

For More Information 47

Index 48

Exploring Stars and Galaxies

Space is full of dazzling clouds of gas like this, which astronomers call nebulas. These clouds are the birthplace of stars.

… INTRODUCTION

Introduction

On every clear night, the twinkling stars shine out of the velvety blackness of space, pouring their faint light onto the darkened Earth. You can see thousands of stars just with your eyes alone. You can see many thousands more when you look at the night sky through binoculars or a telescope.

Telescopes will also show you great bright and dark clouds of gas and dust, which astronomers call nebulas. It is in such clouds that the stars are born. Here and there, telescopes spot stars gathered together in their hundreds of thousands to form dense clusters.

The most powerful telescopes will pick out stars and nebulas and clusters in distant star islands far beyond the stars we see in our own sky. Countless billions of star islands like these, which astronomers call galaxies, make up our universe.

In this book we explore our great universe of stars and galaxies. We discover what stars are like and how they live and die. We look at our home galaxy, to which the Sun and all the other stars in the sky belong, and at other galaxies that shine like beacons in the universe.

Finally, we look at the universe as a whole, finding that it is expanding, as if from a mighty explosion long ago. Astronomers think that a mighty explosion, or Big Bang, did actually happen. They think they know how the universe we know today came about. But they haven't yet solved the final mystery: how will the universe end?

EXPLORING STARS AND GALAXIES

DISTANT SUNS

Located unbelievable distances away in space, stars are found in enormous variety. Some shine feebly, others brilliantly; some are tiny dwarfs, others massive giants; some journey through space alone, others cluster together in their thousands.

To our eyes, the stars look like little points of light. Even in powerful telescopes, which magnify things thousands of times, the stars still appear as points. But this doesn't mean that they are small. What it means is that they are very, very far away from us.

If you could travel through space for many trillions of miles, you would find that the stars are great globes of hot gas that pour out light and heat into space. They are just like the globe of hot gas that appears in our skies every day—the globe we call the Sun. The stars are very distant suns.

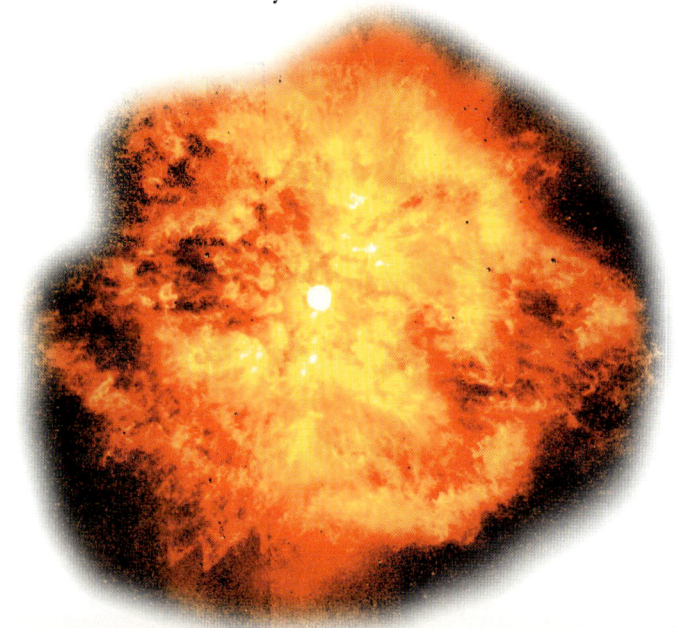

Just how far away are the stars? The answer is: farther than we can ever imagine. Sirius, the brightest star in the sky, is one of the nearest stars to us. But astronomers tell us it is 52 trillion (52,000,000,000,000) miles or 84 trillion kilometers away.

No one can imagine such large figures. The mile or the kilometer is too tiny a unit to measure the vast distances in space.

So astronomers look at distances to the stars in another way—in terms of the time it takes their light to travel to Earth. In a year, light travels nearly 6 trillion miles (10 trillion km).

The light from Sirius takes over 8 years to reach Earth. So astronomers say that the star lies over 8 light-years away. They are using the light-year as a unit to measure distance. Using this unit makes distances in space much easier to understand.

Left: Many stars puff off their outer layers of gas as they die, leaving a tiny white dwarf star behind.

Opposite: Huge globes of stars are found circling around the center of our galaxy. Called globular clusters, they can contain a million or more stars.

Distant Suns

EXPLORING STARS AND GALAXIES

Magnitudes of Brightness

When you look at the night sky, you notice that the stars vary widely in brightness. Some stars stand out like beacons, while others you can hardly see.

Astronomers describe the brightness of a star by its magnitude. A Greek astronomer named Hipparchus first used this method over 2000 years ago. He said that the brightest stars you could see had a brightness of the first magnitude. The dimmest ones had a brightness of the sixth magnitude. Other stars had brightnesses in between.

Bright and dim

Astronomers still use this idea today. But they have extended the scale. They give very bright stars magnitudes smaller than 1 into negative numbers. The brighter the star, the lower the number. The brightest star we see in the sky, Sirius, has a magnitude of –1.45.

Dim stars are given a magnitude greater than 6. And the dimmer, or fainter the star, the higher the number. A faint red star in the constellation Centaurus (Centaur) has a magnitude of about 11. This star, named Proxima Centauri, is famous as the nearest star in the heavens. It lies at a distance of about 4.2 light-years.

True brightness

However, we must remember that how bright a star appears in the sky does not describe how bright it really is. This is because the stars all lie different distances away. So a truly dim star nearby may look brighter than a truly bright star a long way off.

Astronomers can figure out the distance to a few nearby stars by seeing how much the star appears to move against the background of distant stars when viewed from different viewpoints. This is called the parallax method.

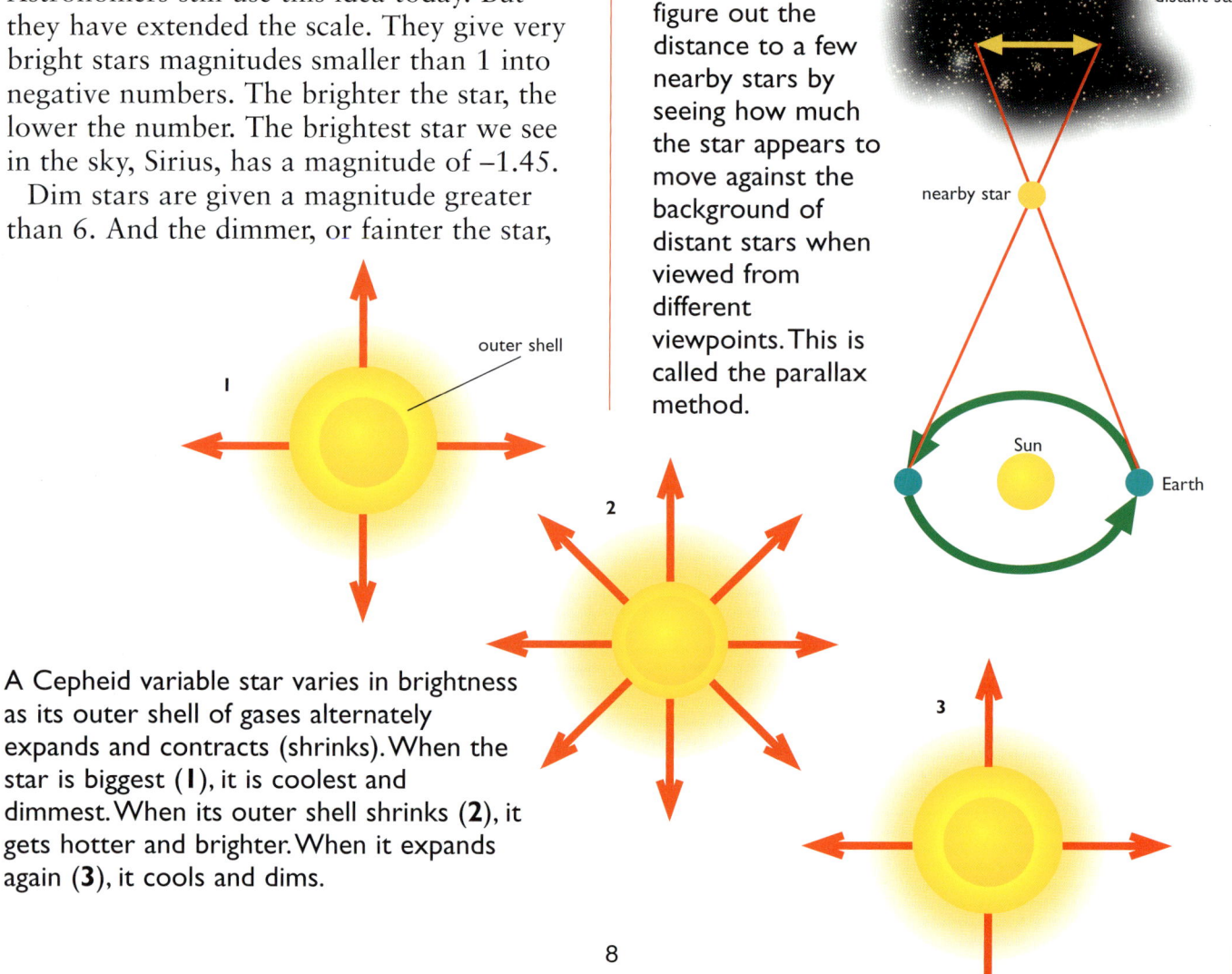

A Cepheid variable star varies in brightness as its outer shell of gases alternately expands and contracts (shrinks). When the star is biggest (**1**), it is coolest and dimmest. When its outer shell shrinks (**2**), it gets hotter and brighter. When it expands again (**3**), it cools and dims.

8

Variable stars

When we look at the stars, they twinkle. It is as though they are flickering like a candle flame and changing in brightness. But most stars shine steadily all the time. They appear to twinkle because of air currents.

However, some stars really do change in brightness. We call them variable stars because their brightness varies over time. Some variables change in brightness because they're actually a pair of stars that circle around each other. As each star regularly passes in front of the other it blocks the other's light. When this happens, the system becomes fainter. The best-known example of this kind of variable is Algol, in the constellation Perseus. It is often called the Winking Demon.

Some stars change in brightness because they change size. These include stars called Cepheids. As they expand, they become dimmer. When they shrink again, they become brighter. American astronomer Henrietta Leavitt pioneered work on the Cepheids early in the 20th century.

Other stars actually explode, and for a while they become thousands or even millions of times brighter than before. Then they slowly fade away.

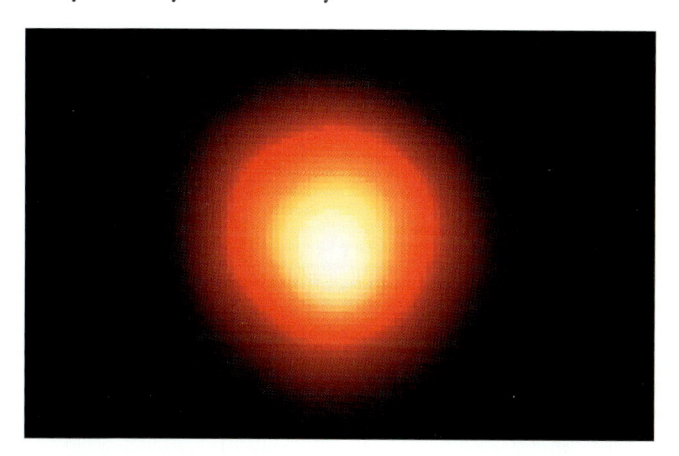

Hubble Space Telescope view of the supergiant star Betelgeuse, in Orion.

Henrietta Leavitt (1868–1921) studied Cepheid variables at Harvard College Observatory in Massachusetts.

Name	Constellation	Apparent magnitude	Distance (light years)
Sirius	Canis Major	−1.45	8.8
Canopus	Carina	−0.73	196
Alpha Centauri	Centaurus	−0.01	4.3
Arcturus	Bootes	−0.06	37
Vega	Lyra	0.04	26
Capella	Auriga	0.08	46
Rigel	Orion	0.11	815
Procyon	Canis Minor	0.35	11.4
Achernar	Eridanus	0.48	127
Beta Centauri	Centaurus	0.60	390
Altair	Aquila	0.77	16
Betelgeuse	Orion	0.80	650
Aldebaran	Taurus	0.85	68
Acrux	Crux	0.9	260
Spica	Virgo	0.96	260
Antares	Scorpius	1.00	425

EXPLORING STARS AND GALAXIES

What Stars Are Like

Stars are great balls of hot gas, with much the same make-up as the Sun. They pour out fantastic amounts of energy into space as light, heat, and other forms of rays.

Stars are made up mainly of two gases—hydrogen and helium. But they contain many other chemical elements as well, such as calcium, iron, and magnesium.

In size, stars vary enormously. With a diameter of 865,000 miles (1,400,000 km), the Sun is a fairly small star. It is classed as a dwarf. Yet other stars can be hundreds of times smaller.

On the other hand, many stars are very much bigger and are termed giants. They can be up to 100 times bigger across than the Sun. But even giants are tiny compared with the biggest stars of all, named supergiants. These stars can be up to 1000 times bigger across than the Sun.

On the move

The stars look as though they are fixed in the night sky. The patterns of stars, the constellations, do not appear to change from century to century. Yet the stars do move—at speeds of up to hundreds of miles a second. Some are hurtling towards us, others are rushing away. The reason why we cannot see them move is because they are so far away.

Only with telescopes can astronomers detect any movement, and then it is only in a few hundred of the nearest stars.

Inside stars

Stars are like great fiery furnaces, pouring out energy into space. Their surface is scorching hot. The temperature of the Sun's surface is about 10,000°F (5,500°C). But the hottest stars are more than ten times hotter.

Blue Giant

Red Giant

Supergiant

Sun

The Sun is a dwarf compared with some other stars. But one day it will expand to become a red giant.

Distant Suns

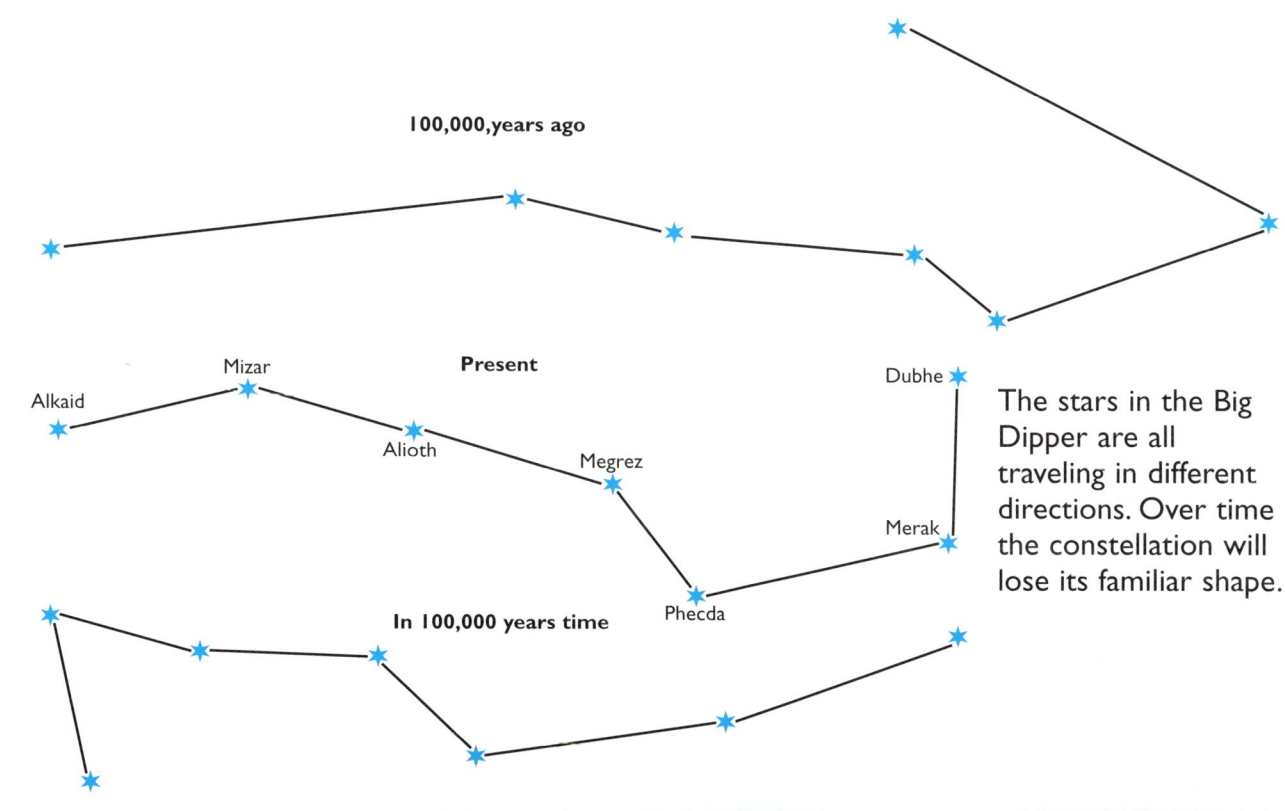

100,000 years ago

Present

Alkaid Mizar Alioth Megrez Dubhe Merak Phecda

In 100,000 years time

The stars in the Big Dipper are all traveling in different directions. Over time the constellation will lose its familiar shape.

The temperatures inside stars are hard to imagine. They rise in the center to 27,000,000°F (15,000,000°C) or more. And the pressure inside stars is millions of times more than the air pressure on Earth.

Under these incredibly high temperatures and pressures, atoms of hydrogen gas are forced to join together. This process is called nuclear fusion. It produces fantastic amounts of energy. And it is this energy that keeps the star shining. The same process is used to produce energy in the hydrogen bomb.

When hydrogen fuses together in nuclear reactions, a new element forms—helium. This explains why helium is the main substance found in stars after hydrogen.

Stars like the Sun use up about 600

A hydrogen bomb explodes when hydrogen atoms are made to fuse (join) together. A similar process powers the stars.

million tons of hydrogen "fuel" every second. Yet they are so big that they can keep shining for billions of years. When their hydrogen runs out, they start to die (see page 24).

Exploring Stars and Galaxies

The white light from the Sun and the stars is actually a mixture of the colors we see displayed in the rainbow.

Looking At Starlight

The stars lie trillions of miles away. Yet astronomers can tell us how hot they are, what they are made of, how fast they are moving, and many more things besides. They gain all this information from the feeble light the stars give out.

They can tell the temperature of the surface of a star from the color of the starlight. The coolest stars are a dull red, hotter ones bright red, and even hotter ones yellow. You can see the same kind of color change as the heating element of an electric stove heats up. It changes color from dull red to yellow as it gets hotter.

The hottest stars of all shine with a bluish-white light.

The colors of the rainbow

To understand how astronomers learn other things about the stars, we must first look at rainbows.

We often see a rainbow when the sun comes out following a rain shower. It is made up of different colors, from violet to red. This happens because sunlight is actually made up

To our eyes, all stars look much the same, shining down on us with a white or yellowish light. But in photographs stars reveal their true colors.

12

Distant Suns

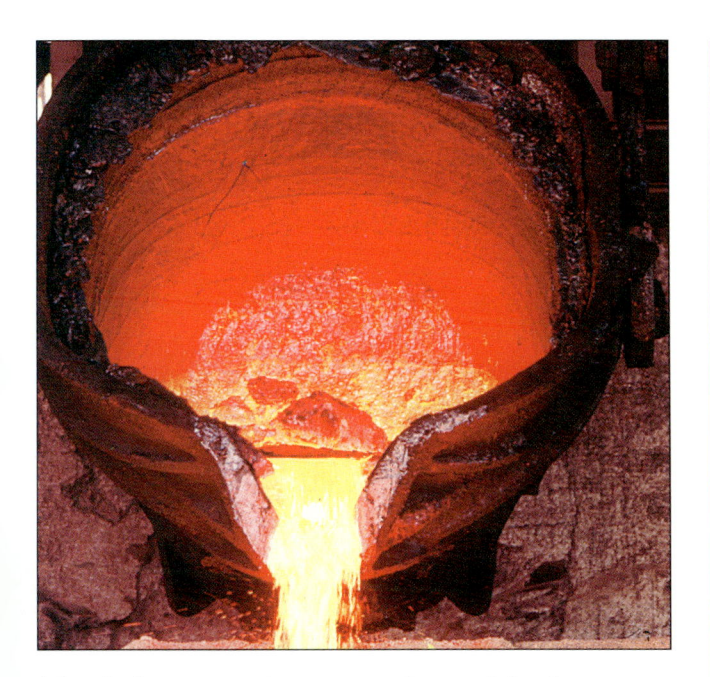

Metal changes color as it is heated. It changes from dull to bright red, then orange, yellow, and white as the temperature goes up. We can estimate the temperature from its color.

of different colors. They make white when they are mixed together. On a drizzly day, sunlight passes through raindrops. They split the light into its different colors, which we see as the rainbow.

We can also split sunlight into its different colors with a wedge of glass, or prism. We call this division of light into color a spectrum. Instruments for dividing light into a spectrum are called spectroscopes.

Tell-tale lines

Astronomers use a spectroscope to get a spectrum from starlight. When they look at a star's spectrum closely, they find that it is crossed by a number of dark lines. These spectral lines provide astronomers with much information.

They can find out about its make-up. This is because certain sets of lines in the spectrum shows that the star contains certain chemical elements.

Left: When sunlight or starlight is passed through a prism, a spectrum of colors is produced.

Below: We can tell a lot about a star from the dark lines in its spectrum. Cool stars (bottom) have more lines in their spectrum than hot stars.

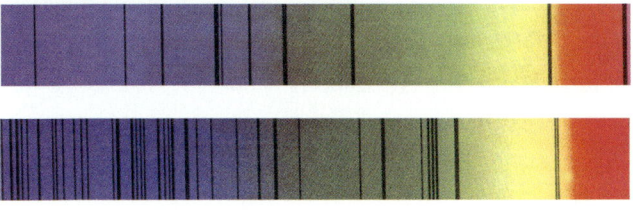

The position of the spectral lines can show the direction a star is moving. If the star is traveling toward us, all the lines move toward the blue end of the spectrum. This is called a blueshift. If the star is traveling away from us, the lines move toward the red end, which is called a redshift.

The same thing happens with sound waves, for example, when an ambulance flashes by. As it races toward us, its siren has a high note (like a blueshift). When it races away, its siren changes to a low note (like a redshift).

13

Exploring Stars and Galaxies

Space shuttle astronauts took this picture of the brightest stars in the constellation Orion. These stars look as if they are grouped together in space. But in fact they all lie at different distances from us.

Traveling Companions

One of the delights of stargazing is looking at double stars—stars that appear close together in the sky. You can see some with the naked eye, including one in the Big Dipper, part of the constellation of Ursa Major (Great Bear). The second star along the handle of the Dipper is Mizar. Close by is a fainter star called Alcor. Both can be clearly seen on a dark night as a pair, or double star.

It looks as if the two stars are close together in space, but they aren't. Alchor is much farther away than Mizar. They appear together in the sky only because they happen to lie in the same direction in space.

A closer look

When you look at Mizar in a small telescope, you find that it is not one star but two. And this time the two stars really are close together in space. They form a two-star system we call a binary.

Astronomers know of many thousands of binary stars. Sirius and Alpha Centauri, two of the brightest stars in the sky, are both binaries.

Eclipsing binaries

In some binaries, the two stars circle round each other like this, as we see them from Earth.

As the stars circle each other, each one regularly passes in front of, then behind, the other. Each time this happens, the brightness of the star system drops. This kind of system is called an eclipsing binary. The best-known example is the star Algol, in the constellation Perseus. Algol dims to one-third of its normal brightness about every 2½ days. This dimming is noticeable to the naked eye, and lasts for a few hours

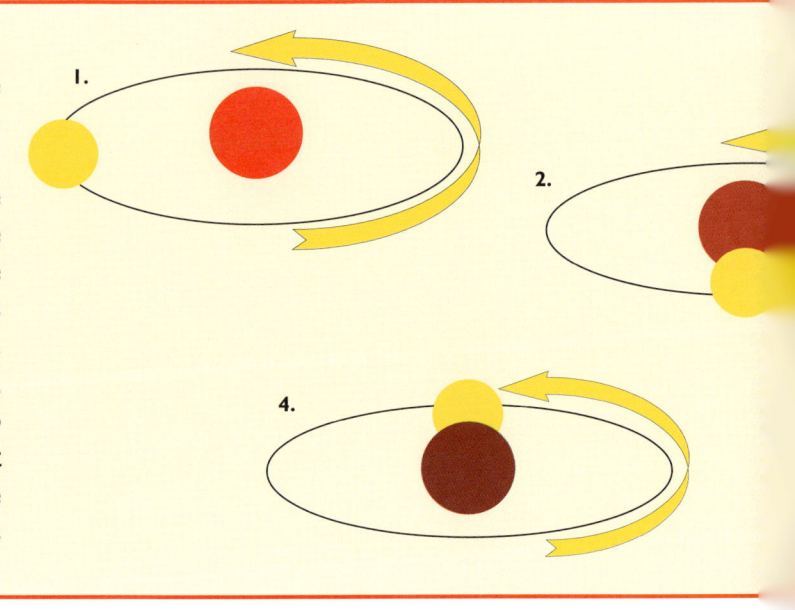

Distant Suns

In some binary systems, the two stars are so close together that we can never see them separately through a telescope. But we can still detect them using a spectroscope.

The two stars in the Mizar binary system each turn out to be binaries when viewed through a spectroscope. This makes Mizar a system of four stars.

Grouping together

Most stars travel through space with companions. Out of every 100 stars, only about 30 stars travel through space alone, like our Sun does.

Some stars travel together in large groups, which we call clusters. In open clusters, the stars are quite widely scattered. We can see one clearly in the Northern Hemisphere, in the constellation Taurus (Bull). It is the Pleiades. It is also called the Seven Sisters for the seven daughters of Atlas, a hero of ancient Greek mythology. Most people, however, can see only six of the seven main stars in the cluster.

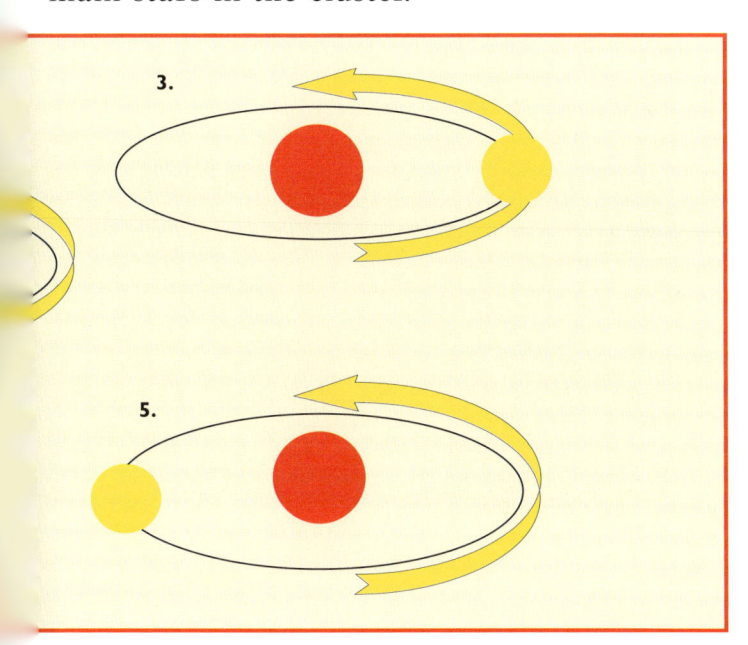

Above: The stars in this fine open cluster sparkle like jewels. It contains several hundred stars altogether.

Below: In a simple binary star system (top), the two stars circle around one another. In this multiple star system (bottom), pairs of stars circle around each other.

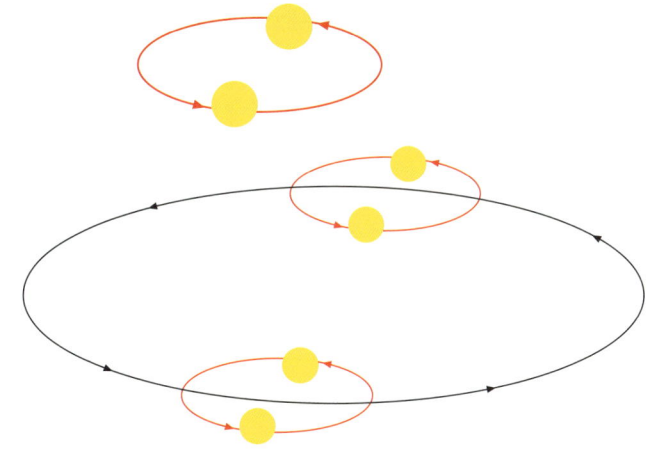

Exploring Stars and Galaxies

A Hubble Space Telescope view of a globular cluster in a neighboring galaxy, the Large Magellanic Cloud. Tens of thousands of stars are on view.

Open clusters

Binoculars and telescopes show that the Pleiades contains many more stars than seven. In all, there are probably as many as 300. They are quite young stars, only about 50 million years old. They are hot and blue-white in color.

There is also another open star cluster in Taurus, around the reddish star Aldebaran, which marks the eye of the bull. It is a V-shaped group called the Hyades.

Globular clusters

In another kind of cluster, however, the stars are packed close together into a globe shape. That is why it is called a globular cluster. In general, the stars in globular clusters are old. In fact they are some of the oldest stars in our star system, or galaxy. Studying them tells us a lot about what our galaxy was like long ago.

In the Northern Hemisphere, there is a fine globular cluster in the constellation Hercules. Known as M13, it contains hundreds of thousands of stars and is just visible to the naked eye as a faint star.

However, the finest globular clusters are to be seen in far southern skies, in the constellations Tucana (Toucan) and Centaurus (Centaur). The two clusters, 47 Tucanae and Omega Centauri, are clearly visible to the naked eye and look magnificent in binoculars.

Clouds in Space

The space between the stars is not quite empty. Particles of gas and minute specks of dust are found scattered everywhere in tiny amounts. We call this stuff interstellar matter.

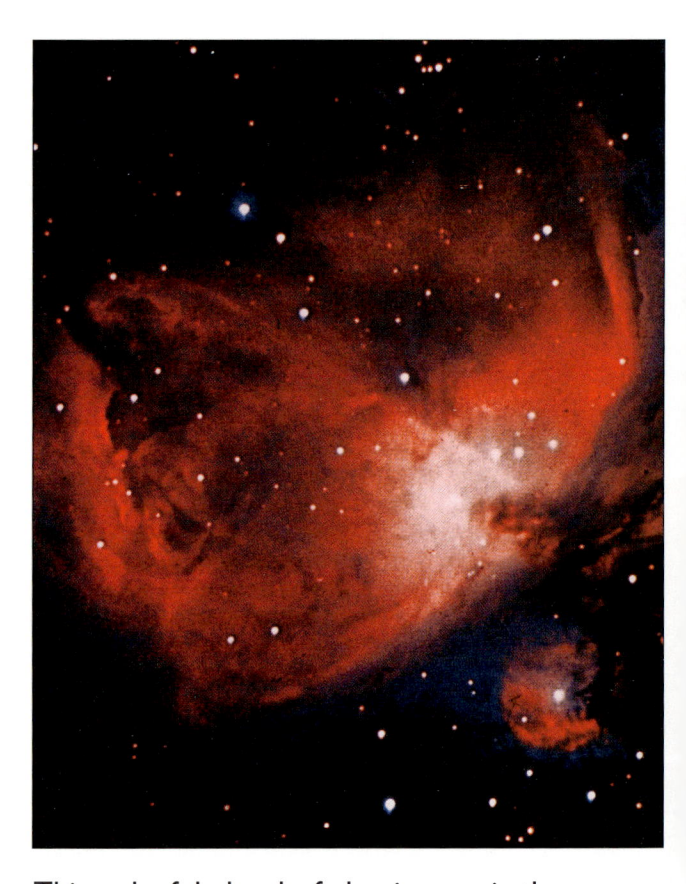

This colorful cloud of glowing gas is the famous Orion Nebula. It is clearly visible to the naked eye.

In places, this matter gathers together to form denser (thicker) clouds, called nebulas (Latin for clouds). Typically, these clouds are about 30 light-years across, but others are much larger, spreading over 100 light-years or more.

Glowing bright

You can see one of the nearest nebulas in the constellation Orion. You see it as a bright patch below the three bright stars that form Orion's Belt.

The Orion Nebula is an example of a bright nebula, which shines. It shines for two reasons. One, its dust reflects the light from nearby stars. Two, its gas particles glow as they give off energy they have taken in from nearby stars.

In the dark

Many nebulas, however, do not shine. We call these dark nebulas. We can see them only when they block the light from stars or bright nebulas behind them. It is the dust in the nebulas that blocks the light. A famous example of a dark nebula is the Horsehead Nebula, also in Orion. It is well-named because it really does look like the head of a horse.

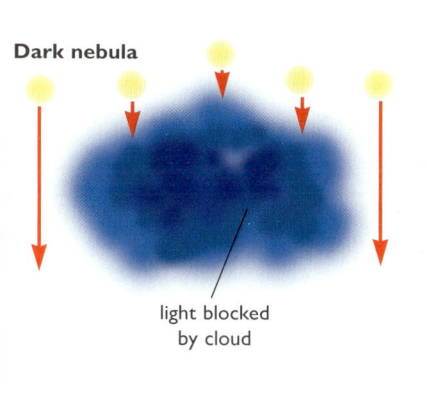

The Trifid Nebula is one of many found in the constellation Sagittarius (Archer).

Below: A dark nebula (left) looks dark because it blots out the light of distant stars. Bright nebulas may shine because they reflect light (center) or because their gas glows (far right).

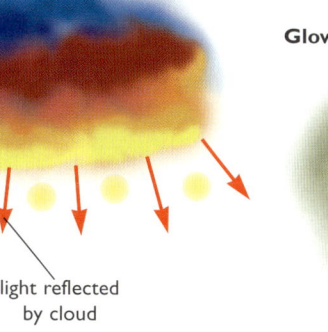

Dark nebula

light blocked by cloud

Reflection nebula

light reflected by cloud

Glowing nebula

stars cause cloud to glow

EXPLORING STARS AND GALAXIES

MATTERS OF LIFE AND DEATH

The stars seem to be everlasting. We see the same ones in the sky today that our ancestors saw thousands of years ago. But stars do change—over periods of millions and billions of years. They are born, they grow up, and they die, just like living things.

Astronomers can never follow all the stages in the life of a single star. This is because stars change only over a very long time—billions of years. But astronomers can study different stars at different stages of their lives. And they can then piece together how a typical star is born, lives, and dies.

In a similar way, we could learn a lot about human lives by studying the people in, say, a shopping mall. We would see babies, toddlers, teenagers, adults, and senior citizens. And over time we would probably figure out how humans live—from babies when they're young to seniors when they're old.

All stars begin their life in a huge cloud of gas and dust. There are many such clouds in the universe. They are made up mainly of hydrogen, with a sprinkling of other chemical elements.

Stars begin to form when parts of the cloud get denser, or thicker. When they get dense enough, the pull of gravity makes its particles begin to attract one another. The cloud begins to collapse, or shrink. As it gets denser, its gravity increases, and it pulls in more and more matter. As gas and dust particles rain down on the shrinking mass, they give up energy. This makes the mass heat up.

By now the mass is spinning around and has formed into a flattened ball shape. The more it shrinks, the faster it spins. This is similar to what happens to ice skaters spinning on the spot. With arms outstretched, they spin slowly, but when they draw in their arms (and shrink), they start to spin faster.

The densest part of the spinning mass is in the middle, and it is this part that may eventually become a star. The remaining matter forms into a thick disc around it.

Matters of Life and Death

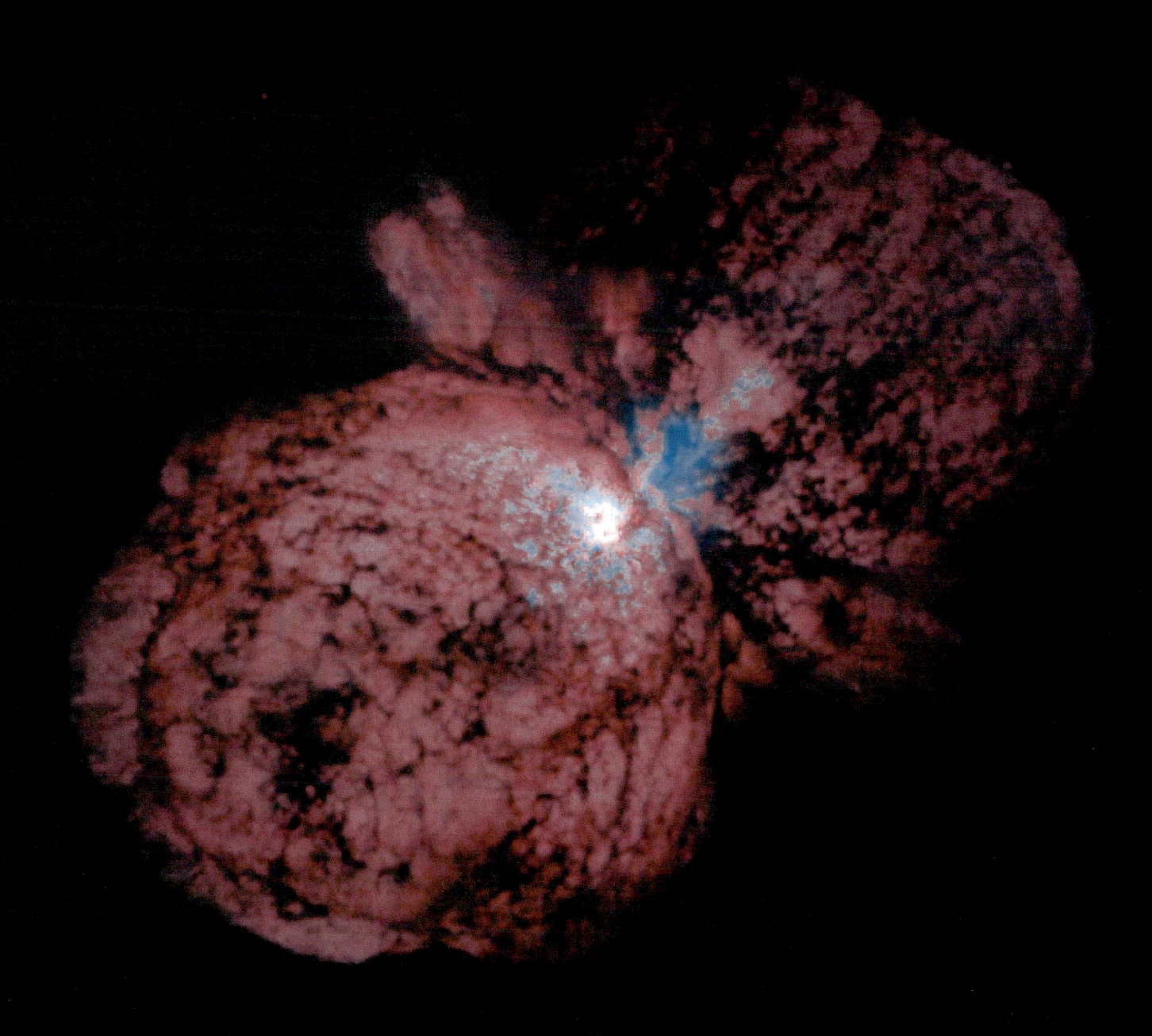

Big stars give off great masses of gas as they approach old age. Eventually, they may blast themselves to pieces.

EXPLORING STARS AND GALAXIES

A Star is Born

Over time, the middle part of the spinning mass of matter gets hotter and hotter and starts to glow. Pressures in the center rise higher and higher.

A dramatic change takes place when the temperature reaches about 20 million degrees F (10 million degrees C). Nuclear reactions begin: hydrogen atoms smash into one another and fuse together (see page 11). Enormous energy is produced, which pours out into space as light, heat, and other radiation. The shining body is a new star.

For a while, the newborn star continues shrinking. But soon the pressure of the energy pouring outward from the core balances the pressure of the matter pulled inward by gravity. The star reaches a steady state, staying the same size and having the same energy output, or brightness.

The lives of the stars

How long a star shines steadily like this depends mainly on its mass. The smaller the mass a star has, the longer it will live.

The smallest stars have a mass only about one-tenth of the Sun's mass. They are much cooler than the Sun and give off reddish

MATTERS OF LIFE AND DEATH

light. That is why we call them red dwarfs. These bodies live the longest of all stars. Some can probably live for as long as 200 billion years. This is more than ten times as old as the universe is now!

Like the Sun

Our own star, the Sun, has been shining steadily for about 5 billion years. Astronomers calculate that it will probably carry on shining steadily for another 5 billion years. Then, it will start to die. Stars with a similar mass to the Sun will probably shine steadily for about the same length of time before they die.

> **Brown dwarfs**
> Not all the masses of matter that form out of the gas and dust clouds become stars. Some don't grow big and heavy enough. And this means that the temperatures and pressures inside them can't rise high enough to trigger off nuclear reactions. These failed stars remain warm, glowing balls, known as brown dwarfs.
>
> Astronomers think that brown dwarfs may be similar in make-up to the planet Jupiter, but with up to 100 times more mass. They probably have a thick atmosphere of hydrogen above a deep liquid hydrogen ocean. The Hubble Space Telescope has spotted many objects that could be brown dwarfs.

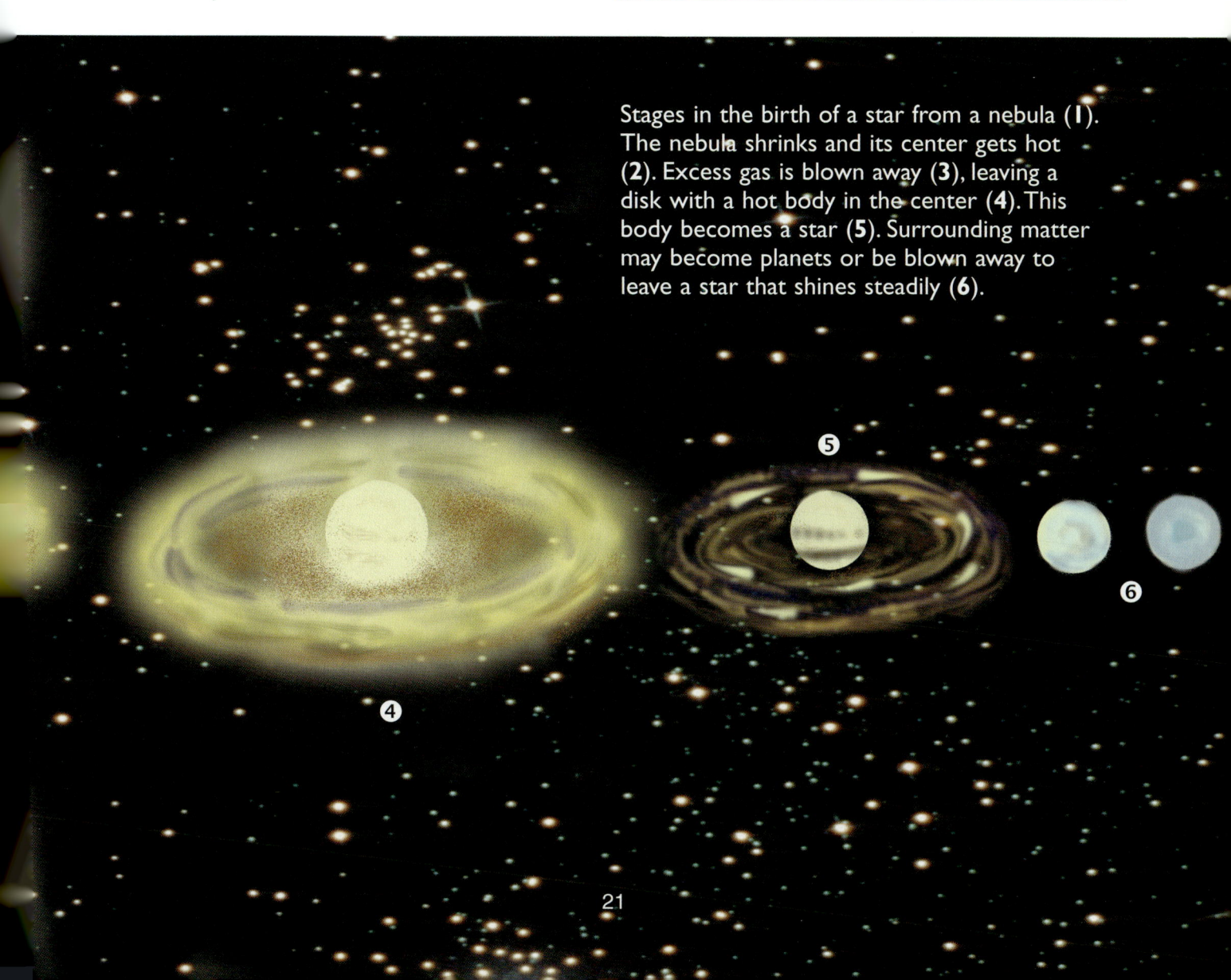

Stages in the birth of a star from a nebula (**1**). The nebula shrinks and its center gets hot (**2**). Excess gas is blown away (**3**), leaving a disk with a hot body in the center (**4**). This body becomes a star (**5**). Surrounding matter may become planets or be blown away to leave a star that shines steadily (**6**).

EXPLORING STARS AND GALAXIES

Sun-like Stars

A star with a similar mass to the Sun shines steadily for about 10 billion years. During this time, it is "burning" up the hydrogen fuel in its center, or core, in nuclear reactions to make helium.

When all the hydrogen is used up, the reactions stop. Gravity causes the core to contract, or shrink. As it shrinks, it heats up. The energy it gives off makes the outer part of the star swell up, or expand.

Over time, the star can grow up to 100 times bigger across, turning into a red giant. You can get the picture if you imagine an apple seed growing into something the size of a watermelon. When the Sun becomes a red giant, its outer layers will expand out beyond the planet Mercury, and Earth will become hotter than an oven.

The core gets hotter and hotter as it shrinks. Eventually it becomes so hot that it triggers off more nuclear reactions, this time between the helium atoms. In these reactions, the helium changes into carbon. When the helium is used up, the reactions end. The core and the rest of the star begin to contract with the force of gravity.

Smoke rings in space

As the dying star shrinks, it regularly puffs off clouds of gas. They form a kind of shell around the star, which expands outward as time goes by. The shell of gas is lit up by the energy the shrinking star gives out.

The core of the star grows smaller and

How a star like the Sun dies. After shining steadily for ages (1), it starts to expand (2, 3, 4) until it becomes a red giant (5). Over time the red giant shrinks, puffing off gas as it does so (6). All that remains is a tiny white dwarf star (7).

smaller, growing ever brighter. It turns into a star we call a white dwarf.

We can see many dying stars like this in the heavens, surrounded by a glowing shell of gas. Early astronomers called them planetary nebulas because in telescopes they appear as a disc, like the planets. But planetary nebulas have nothing to do with the real planets.

A classic planetary nebula is the Ring Nebula in the constellation Lyra (Lyre). It looks like a colorful smoke ring.

Heavy dwarf

White dwarfs are white and hot. But they do not look very bright to us because they are so small. Most are only about the size of Earth. But they can contain as much mass as the Sun. So they are very heavy for their size. A cupful of their matter would probably weigh as much as 100 tons on Earth.

One of the best-known white dwarfs is a companion of Sirius, brightest star in the sky. It was also the first white dwarf to be discovered. Sirius is called the Dog Star, so its companion is often called the Pup.

The end

Over time, white dwarfs gradually cool down as their energy disperses into space. They will become dimmer and dimmer. One day they will darken and turn into black dwarfs and disappear from sight. No one knows how long this takes. Maybe the universe is not old enough for black dwarfs to have formed yet.

EXPLORING STARS AND GALAXIES

The Death of Massive Stars

Some stars are much bigger than the Sun, more than ten times the Sun's mass. These heavyweight stars burn their hydrogen fuel quickly. And they shine thousands of times more brilliantly than the Sun. But they have a much shorter life. They may shine steadily for only a few million years old before they use up their hydrogen and start to die.

As in smaller stars, the core of a big star begins to collapse and heat up when it starts to die. And its outer layers expand. In time it grows into a huge supergiant star many hundreds of times wider across than the Sun.

The temperature of the massive core of the supergiant rises to hundreds of millions of degrees. And it becomes hot enough for many other kinds of nuclear reactions to take place, which produce other chemical elements. When no more reactions can take place, the core collapses under its own weight—in just a few seconds.

Left: Wisps of glowing gas from an ancient supernova explosion. They travel through space for many light years.

MATTERS OF LIFE AND DEATH

Going supernova

The energy released by the collapsing core sets off a fantastic explosion, called a supernova. The star blasts itself apart, flaring up to become billions of times brighter than it was. For a while it may become as bright as an entire galaxy of stars.

Over the past 1,000 years astronomers have recorded four supernovas in our galaxy. Chinese astronomers spotted one in 1054. They called it a ghost star. Skilled observers, they noted where it was in the heavens and that it was so bright that it could be seen in the daytime.

We see the remains of the 1054 supernova today as a cloud of gas in the constellation Taurus (Bull). We call it the Crab Nebula.

Below: How a massive star dies. After shining steadily (**1**), it swells up (**2**, **3**, **4**) until it becomes a supergiant (**5**). Then it explodes as a supernova (**6**).

Exploring Stars and Galaxies

began observing what remained of the explosion—slowly expanding rings of glowing gas.

Seeding the universe

Supernovas scatter into space all the chemical elements that made up the original star. These elements—nature's building blocks—become part of the great clouds, or nebulas, in which new stars will be born. The Sun and the planets were born in such a cloud. They contain atoms and elements made in the nuclear furnace of a long-dead star. So do we.

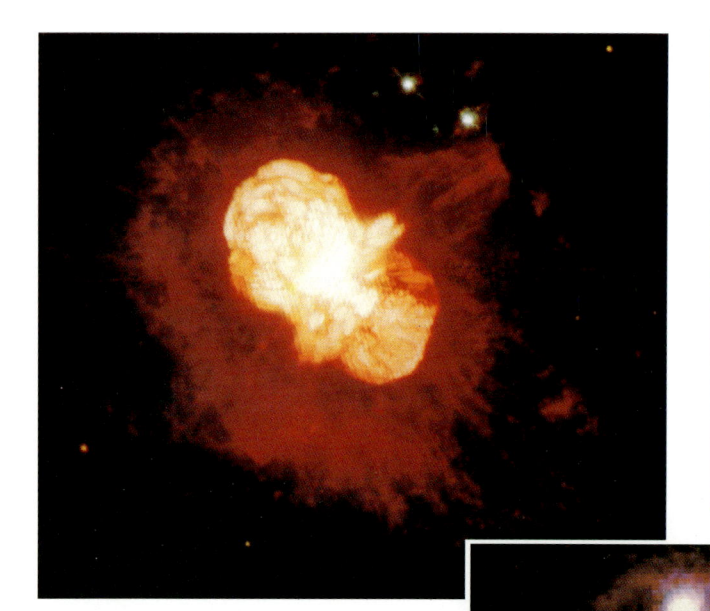

Above: An expanding shell of gas hides one of the biggest stars known, named Eta Carinae. This star will probably not survive for long before it becomes a supernova.

Left: Rings of glowing gas surround what is left of the star that became a supernova in 1987.

Below: These delicate wisps of gas from a supernova that took place in the constellation Cygnus (Swan) form the Veil Nebula.

Supernova 1987A

Supernovas are so bright that they can be spotted not only in our own galaxy but in other galaxies as well. In February 1987, a supernova erupted in the nearest galaxy to our own, the Large Magellanic Cloud. It was so bright that it could easily be seen with the naked eye. Astronomers reckoned that it was about 250 million times brighter than the Sun.

Supernova 1987A was the brightest exploding star seen in our skies since 1604—before the telescope was invented. This time astronomers around the world trained their telescopes on the star to follow the progress of the explosion. A few years later the Hubble Space Telescope

Crushed

All that remains of a massive star after it has exploded as a supernova is the collapsing core. What happens next depends on the core's mass.

If its mass is up to about three times the mass of the Sun, the core shrinks to form a tiny body called a neutron star. Here is what happens.

As the core collapses under its own weight, it gets smaller and smaller. The smaller it gets, the stronger its gravity becomes, and the smaller and denser it becomes. In only seconds, it shrinks to a body only about 15 miles (25 km) across. Its atoms have been crushed, and it is made up mainly of tiny particles called neutrons. We now call it a neutron star.

Superdense

A neutron star is an amazing body—it has the mass of the Sun squeezed into a ball the size of a city. It is therefore incredibly dense —just a pinhead of its matter would weigh a million tons.

A neutron star also spins around rapidly. Its powerful magnetic field spins with it. Particles in the space around the star get caught up in the spinning magnetic field and whirled around. They eventually escape into space as beams of radiation.

The heavenly lighthouse

Because a neutron star spins, its beams sweep around in space like the beams from a lighthouse. From Earth, we observe them as flashes, or pulses, of radiation as they sweep past us. These pulses are as regular as the ticking of a clock.

MATTERS OF LIFE AND DEATH

Left: As a neutron star spins around, its beams sweep around, too.

Below: The beams from a rotating neutron star sweep through space (**1**). We see a pulse as it sweeps past Earth (**2**), before continuing on its way (**3**).

Exploring Stars and Galaxies

Little Green Men
Radio astronomer Jocelyn Bell at Cambridge, England, discovered the first pulsing signals from space in 1967. At first no one knew what caused them. It was even suggested that they were signals sent by intelligent beings, so they wre dubbed LGMs, standing for Little Green Men.

Soon many more pulsating sources, or pulsars, were detected, and it was realized that they were rapidly spinning neutron stars.

Hundreds of pulsars have now been discovered. Many spin around relatively slowly, about once a second. But some spin around more than 600 times a second. Just think of it—a body the size of a large city spinning around hundreds of times in a second.

Into a hole
A very big star doesn't end up as a neutron star. If the mass of its core is more than about three times the mass of the Sun, it will become an object called a black hole.

In the very heavy core of the big star, gravity is incredibly strong. It pulls all the matter in the core together with such force that it gets crushed.

Astronomers believe that the core eventually gets crushed into a tiny point. All that remains is a region around it that has a fantastically powerful gravitational pull.

This picture shows a galaxy known as M87. A jet of material shoots out from a bright mass of gas (inset), which could hide a black hole.

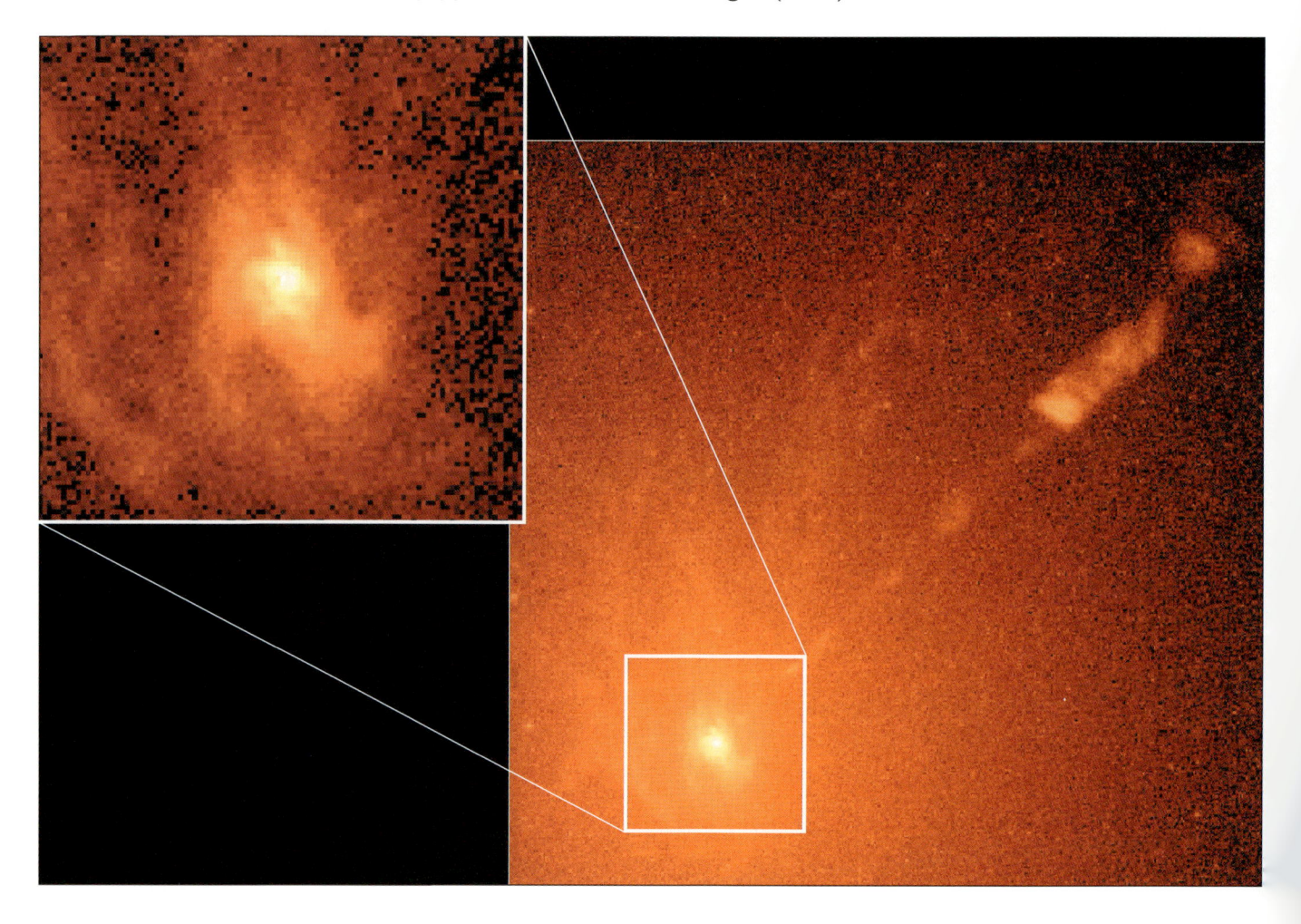

MATTERS OF LIFE AND DEATH

Anything that strays into this region is swallowed up—even light rays. It is because it swallows light that such a region is called a black hole.

Discovering black holes

Because light can't escape from a black hole, we can't see it. But we can detect it in other ways. Like the star it came from, the black hole spins around. Nearby matter attracted by the black hole's gravity is dragged around too before being sucked inside. It forms a swirling disk of matter.

As it swirls furiously around, this matter gets incredibly hot, reaching temperatures of 200 million degrees F (100 million degrees C) or more. This makes it give off energy as X-rays, which can be detected from Earth. Astronomers have picked up a powerful source of X-rays in the constellation Cygnus, called Cygnus X-1. They think that it comes from a swirling disk around a black hole.

The Cygnus X-1 black hole is part of a binary, or two-star, system. The other star is a blue supergiant. Astronomers think that the black hole is gradually pulling matter from this star into it. They reckon that this black hole is about 35 miles (60 km) across and has a mass about 10 times the mass of the Sun.

When there is a black hole in a binary system, it gradually draws off matter from the other star. The matter forms a fast spinning disk before being swallowed up.

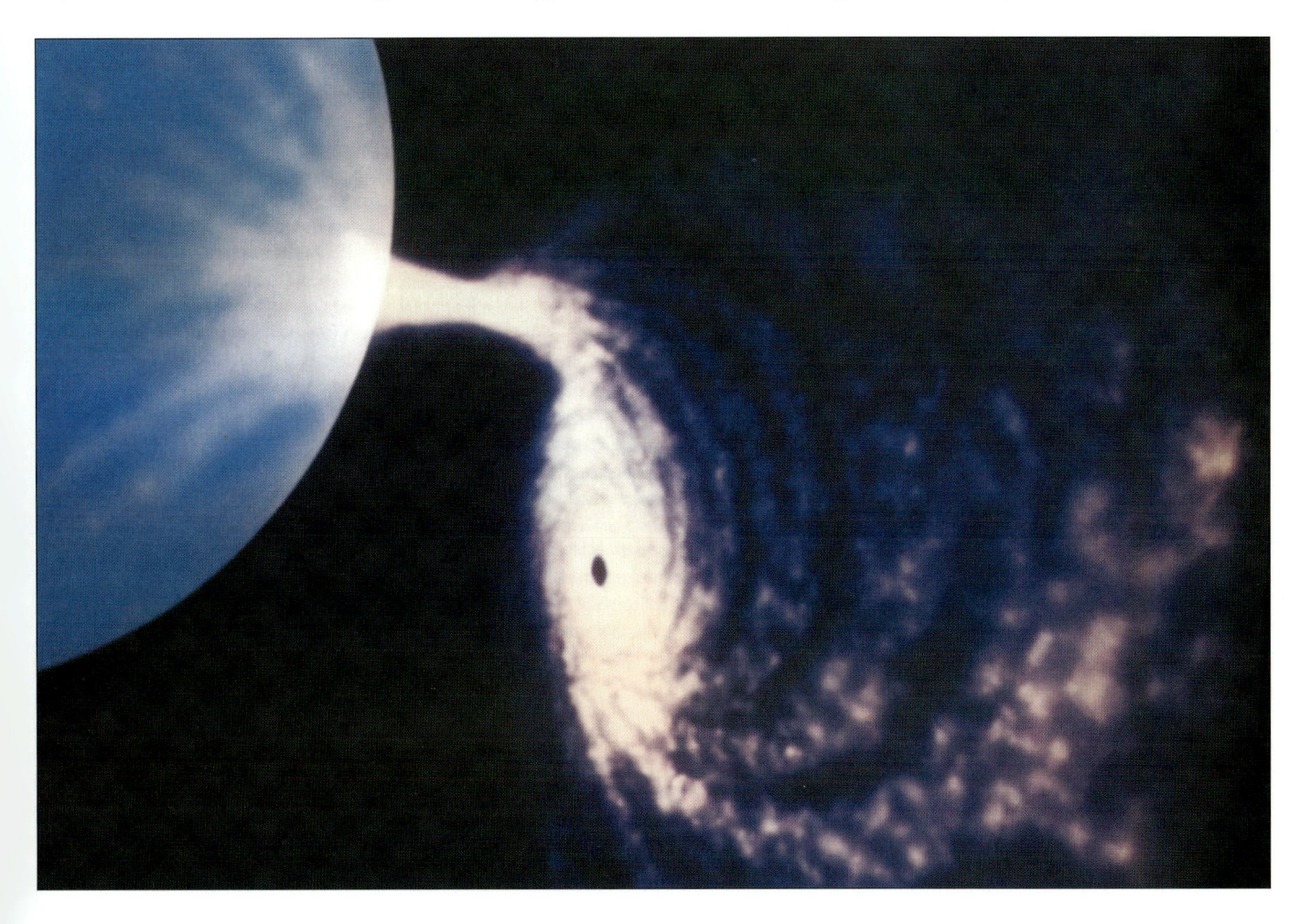

EXPLORING STARS AND GALAXIES

GALAXIES OF STARS

The stars are not scattered evenly throughout space. They are found clustered together in great star "islands," with empty space in between. These star "islands," or galaxies, contain billions of stars and measure thousands of light-years across.

In powerful telescopes, astronomers can see galaxies in almost every direction they look in space. In some parts of the sky they can see hundreds, even thousands of galaxies gathered together in clusters. Billions of galaxies make up the universe.

Most galaxies lie very far away. In small telescopes, they look like fuzzy blobs, rather like the gas and dust clouds we call nebulas. And astronomers once thought that they were nebulas. As telescopes improved, many of these "nebulas" were seen to have a spiral shape. Some astronomers thought that these spirals were located in our own galaxy, while others thought that they might be located outside it.

In 1919, a US astronomer named Edwin Hubble began investigating the spiral nebulas at the Mount Wilson

Left: Edwin Hubble

Observatory outside Los Angeles. He used what was then the world's most powerful telescope, the Hooker. It had a light-gathering mirror 100 inches (2.5 meters) in diameter.

In this telescope, Hubble saw for the first time individual stars in the large spiral nebula in Andromeda. Among them he found variable stars called Cepheids. From the way the brightness of Cepheids changes, astronomers can figure out how far away they are.

Hubble worked out that the Cepheids he spotted in the nebula in Andromeda were nearly 1 million light-years away. This put them far beyond our own galaxy. The nebula had to be a separate star system—a separate galaxy. Since then, astronomers have found that the Andromeda Galaxy is more than twice as far away as Hubble calculated, at a distance of some 2.3 million light-years.

Opposite: Spiral galaxies like this are found throughout space. They can contain hundreds of billions of stars and measure hundreds of thousands of light-years across.

Galaxies of Stars

EXPLORING STARS AND GALAXIES

Classifying the Galaxies

Edwin Hubble found that galaxies came in many different sizes but usually had similar basic shapes. So he decided to classify galaxies by their shape, and we still follow his method today. There are two main kinds of galaxies—spirals and ellipticals. The others have no definite shape and are known as irregulars.

Maybe as many as three-fourths of all galaxies are spirals. They are shaped rather like pinwheels. They have a central bulge of stars, surrounded by a disk of stars, gas, and dust. Within the disk, most of the stars are young and lie mainly on curved "arms" that give the galaxy its spiral shape.

In some spiral galaxies, the arms are close together, while in others they are quite open. Some galaxies have a bar of stars through the central bulge and are known as barred-spiral galaxies.

In a spin

Spiral galaxies spin around bodily in space, like a whirling pinwheel. You can tell in which direction they spin by looking at the curved arms. They trail behind the direction of spin like the streams of water from a garden sprinkler.

The stars in the disk circle within the disk region. It takes them millions of years to complete a full circle. But the stars within the bulge at the center of a galaxy circle in many directions. Great globe-shaped clusters of older stars circle outside the bulge. These globular clusters travel within a huge sphere around the whole galaxy, called the halo.

Galaxies can have many shapes. Spherical (**1**) and egg-shaped (**2**) ones are classed as ellipticals. Others can be ordinary spirals (**3**) or barred spirals (**4**).

32

GALAXIES OF STARS

Elliptical galaxies

Elliptical galaxies can be round or oval in shape. They don't have the curving arms of spirals. Also they contain no dust and have no new stars forming in them. They are made up mainly of older stars.

The biggest galaxies in the universe are ellipticals. They are found in the centers of the great clusters of galaxies that exist in space. They probably got so large by gobbling up smaller galaxies that strayed too close to them.

When galaxies collide

Astronomers have found plenty of evidence that galaxies periodically bump into one another. When galaxies collide, their stars don't generally smash into one another. They pass each other by like ships in the night. This is because galaxies are mostly empty space, with the stars scattered about with plenty of space between them.

The Hubble Space Telescope provided this head-on view of galaxy NGC 4013 (above) and two galaxies colliding (below).

EXPLORING STARS AND GALAXIES

The Milky Way

On a clear moonless night, you can often see a faint, misty band of light arching across the sky. We call it the Milky Way. When you look at it in binoculars or a telescope, you see that it is made up of countless stars, seemingly packed close together.

What is the Milky Way? It is a view of our own galaxy from the inside. That is why we also call our galaxy the Milky Way, or sometimes just the Galaxy.

Where the Sun is

The Sun and all the other stars we see in the sky belong to the Milky Way Galaxy. In all, the Milky Way contains something like 100 billion stars. It seems to be a typical spiral galaxy, with a central bulge and a surrounding disk. It seems to have a faint bar through the center, so maybe it is a barred-spiral galaxy.

The Milky Way Galaxy measures about 100,000 light-years across. The Sun sits on one of the curved arms that come out of the bulge, about 25,000 light-years from the center.

In a halo

As in other spirals, globular clusters circle outside the bulge in a spherical halo that surrounds the whole galaxy. Nearly 150 of these clusters are known, the largest containing several million old stars.

The Milky Way rotates slowly in space. The stars take different periods of time to circle around the center, depending on how far out they are. The Sun, for example, takes about 225 million years to circle once around the galaxy.

disk

bulge

halo

globular clusters

Above: Our galaxy has a bulge of stars in the center and other stars on curved arms in the disk. All around it is a great sphere, or halo, of faint matter.

Left: This picture of our galaxy was produced from images sent back by a satellite.

Galaxies of Stars

Galactic Neighbors

Most galaxies lie so far away that we can see them only in telescopes. But there are three that we can see with the naked eye.

In the far southern hemisphere, you can spot two faint misty patches. The larger is called the Large Magellanic Cloud (LMC), the smaller, the Small Magellanic Cloud (SMC). They are named after the Portuguese navigator Ferdinand Magellan. He led the first expedition to sail around the world, setting out in 1519. He sailed across the southern oceans, from where the two Clouds could be seen.

Telescopes show that the LMC and SMC Clouds are really separate galaxies. They are in fact the nearest galaxies to the Milky Way. The LMC is closest, at a distance of only 170,000 light-years. It is a small galaxy, with only about one-third of the diameter of our own galaxy. The SMC is only about half the size of the LMC.

The great spiral

The other galaxy that we can see with the naked eye lies in the northern hemisphere in the constellation Andromeda. It was once called the Great Spiral in Andromeda, and great it is. It is one-and-a-half times bigger across than our own galaxy and it contains billions more stars.

The only reason we can see the Andromeda Galaxy in the night sky is because it is so large, for it lies an incredible 2.3 million light-years away. It the farthest object that we can see with the naked eye.

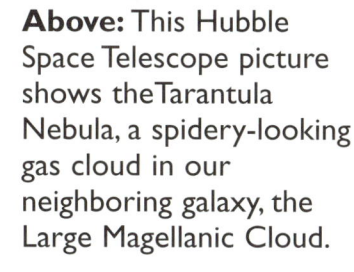

Above: This Hubble Space Telescope picture shows the Tarantula Nebula, a spidery-looking gas cloud in our neighboring galaxy, the Large Magellanic Cloud.

Left: The Andromeda Galaxy is a spiral like our own galaxy, but it is much bigger. Close by are two smaller elliptical galaxies, which are linked to it by gravity.

Exploring Stars and Galaxies

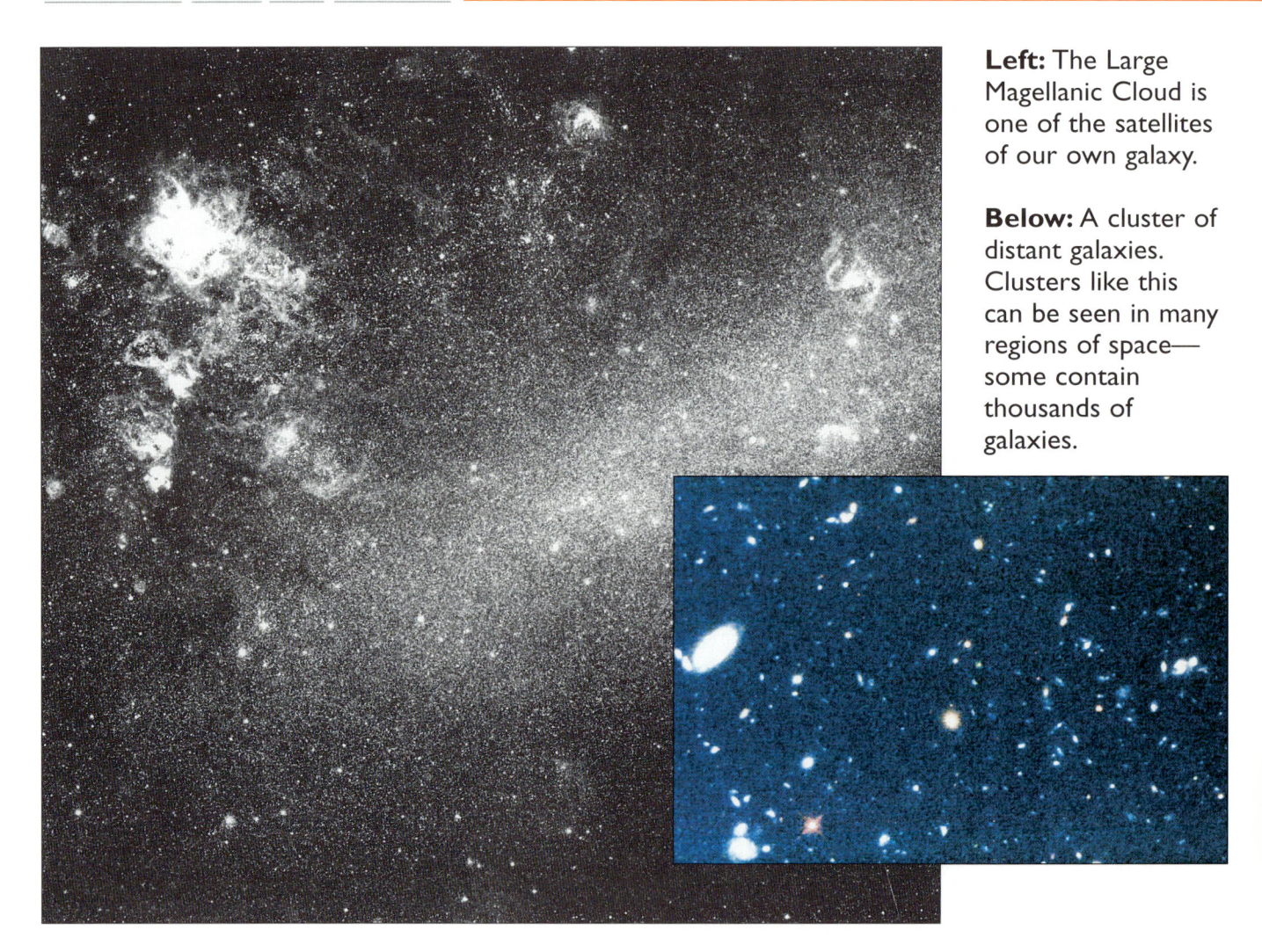

Left: The Large Magellanic Cloud is one of the satellites of our own galaxy.

Below: A cluster of distant galaxies. Clusters like this can be seen in many regions of space—some contain thousands of galaxies.

Clustering Together

The Magellanic Clouds, the Andromeda Galaxy, and the Milky Way are neighbors in space. They form part of a cluster of galaxies in our part of the universe known as the Local Group. All the galaxies in the Group are bound together by gravity.

The Magellanic Clouds circle around our galaxy like space satellites around Earth, and they are known as satellite galaxies. They are gradually coming closer, and one day our galaxy will swallow them up. The Andromeda galaxy has smaller satellite galaxies too.

Scattered about

In all there are over 30 galaxies in the Local Group, and they are scattered over a region of space about 7 million light-years across. The Andromeda Galaxy is the biggest, followed by the Milky Way; then comes M33, in the constellation Triangulum (Triangle). These three galaxies are the only spirals. The rest are small elliptical or irregular galaxies.

The more distant galaxies gather together in clusters too. Some of these clusters contain thousands of individual galaxies. An example is the cluster in the constellation Virgo, which contains at least 1,000 galaxies.

36

GALAXIES OF STARS

Left: A distant quasar, which pours out fantastic energy into space as light and other kinds of radiation.

Active Galaxies

Most galaxies give off much of their energy as light. The amount of energy they give off is what would be expected for a group of billions of stars. But a few galaxies give off much more energy than expected. We call them active galaxies.

Some active galaxies pump out their exceptional energy as light. Others pour it out most of their energy in the form of radio waves, or X-rays.

Amazing quasars

Astronomers also pick up powerful radio beams from incredibly distant objects called quasars. These objects lie billions of light-years away, yet are still visible as faint stars. They seem to produce as much energy as a thousand normal galaxies.

Where do active galaxies get their energy? Certainly, the ordinary nuclear processes in stars couldn't provide it. So astronomers think that the energy comes from black holes in the heart of the galaxies. A black hole produces energy from the ring of matter swirling round it.

The black holes in active galaxies would have to have millions of times the mass of our Sun to produce the amazing energy given out.

Left: This donut-shaped ring of dusty matter is the center of a giant active galaxy. Astronomers think that a black hole lurks inside it.

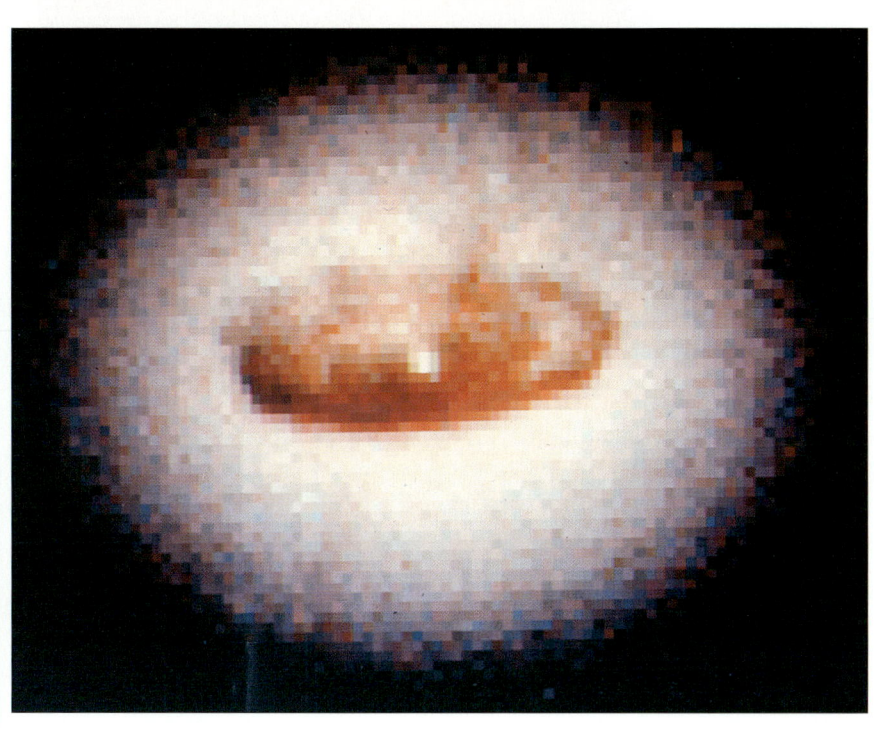

EXPLORING STARS AND GALAXIES

THE MIGHTY UNIVERSE

All the planets and stars, the galaxies and clusters, and space itself make up the universe. Astronomers think they know what the universe is like and how it began in a huge explosion long ago. But they're not sure what will happen to it in the future.

All things that exist—rocks, air, living things, and the heavenly bodies floating in space—make up the universe. Another name for the universe is the cosmos.

The universe as we know it today is made up mainly of bodies floating in space—planets and their moons, stars, and galaxies. These bodies are widely scattered, and most of the universe is just empty space.

In early times, people had no idea what the universe was like. They thought that our world was the center of the universe, and had different ideas about what it was like. In ancient India, for example, people thought that the world was carried on the back of four elephants. In turn, the elephants stood on the shell of a huge tortoise. And the tortoise itself was carried by a snake swimming in a vast ocean.

The ancient Greeks knew Earth was round. They believed it lay in the center of a great celestial, or heavenly, sphere, which spun around. The stars were stuck to the inside of the sphere. The Sun, the Moon and the planets circled Earth closer in. The celestial sphere was the universe.

In 1543 an astronomer named Copernicus turned ideas on their head by suggesting that Earth circled around the Sun, and not the other way round. The other planets circled round the Sun as well, forming a family of bodies called the solar system. The universe then seemed to be the solar system.

Over time, astronomers realized that the Sun was merely a star like the other stars in the night sky. Then the universe became the great star system we call our galaxy.

This takes us to the beginning of last century. Then astronomers found that there were many more galaxies around. So the universe seemed to be made up of galaxies and clusters of galaxies floating in space. And that is more or less what astronomers believe today.

Opposite: The universe is bigger than we can ever imagine. It is mostly empty space, with galaxies of stars, dust, and gas scattered here and there.

THE MIGHTY UNIVERSE

Exploring Stars and Galaxies

1

Astronomers think that a violent event they call the Big Bang (**1**) created the universe and began its expansion. For a while it was incredibly hot (**2**).

The Expanding Universe

We saw earlier that astronomers can tell how a star is moving by studying the light it gives out. They look at the way the dark lines in its spectrum shift towards the blue end or the red end (see page 13). A blueshift shows that the star is traveling toward us, a redshift that the star is traveling away.

Astronomers can tell whether a galaxy is traveling toward or away from us in a similar way. When they study the light from galaxies, they find that it almost always shows a redshift. In other words, nearly all the galaxies are rushing away from us. And the farther they are away, the faster they seem to be traveling.

It seems as if the whole universe is getting bigger, or expanding. If this is true, the universe must have been smaller in the past. Working backward, there must have been a time when everything in the universe was packed together in one place. Then something must have happened that set the universe expanding.

Most astronomers agree that something like this actually happened. The universe came into being as a point, and a great explosion began its expansion. They call this explosion the Big Bang. It probably happened about 15 billion years ago, but astronomers don't know exactly when.

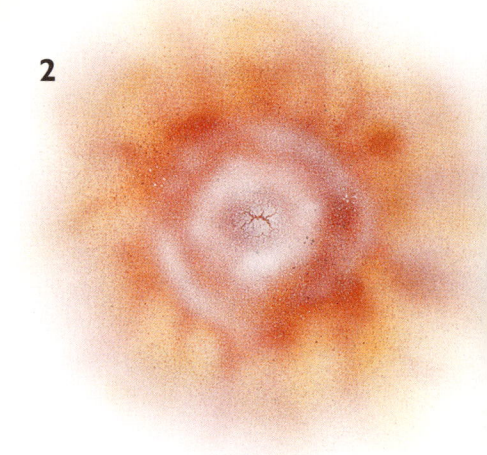

2

After the bang

The astronomers who study the universe as a whole and how it has developed are known as cosmologists. Incredibly, they think that they know what has happened to the universe nearly from the instant it was born.

To start with, the universe was tiny and incredibly hot—with temperatures of trillions of trillions of degrees. It consisted only of energy, because no matter could exist at such fantastic temperatures. But as the universe expanded rapidly, it also cooled down rapidly.

As it cooled down, tiny bits of matter—particles—began to form. Over time, these particles joined together to form simple atoms of hydrogen. And later the atoms formed into clouds. It was from these clouds that the first stars were born. In time, the

THE MIGHTY UNIVERSE

stars began to collect together into galaxies. No one is certain when this took place. But it may have happened before the universe was 1 billion years old.

The Universe Today

Fifteen billion years after the Big Bang, we come to the universe as we find it today. It has expanded to become bigger than we can ever imagine. The farthest objects we can see in our most powerful telescopes seem to lie nearly 15 billion light-years away. (In miles this is 10 followed by 22 zeros, or in kilometers 16 followed by 22 zeros.)

Looking at things in another way, the light from these objects has been traveling toward us for nearly 15 billion years. So we are seeing them as they were shortly after the Big Bang, not as they are today.

Above: We find galaxies everywhere when we look deep into space. The farthest ones lie more than 10 billion light-years away.

Then, as the universe got bigger, it cooled down, and matter started to form (**3**). Some 15 billion years after the Big Bang, we find the universe as it is today (**4**), still expanding.

Exploring Stars and Galaxies

Structure of the Universe

The diagram on the previous page reminds us of the structure of the universe as we find it today, from planets and stars, galaxies and clusters, to enormous superclusters. Curiously, the superclusters seem to form kinds of thin sheets around otherwise empty spaces, called voids.

The End

We think we know what the universe was like in the very distant past, and what it is like today. But what will happen to it in the very distant future?

There seem to be three main possibilities. One is that the universe will continue to expand forever until it eventually runs out of energy. Another idea is that the universe will eventually stop expanding, leaving it at a certain size.

The third idea is that the expansion will stop expanding and then go into reverse—it will start to shrink. It will carry on shrinking until it ends up, as it began, as a tiny point. Astronomers say there would be a Big Crunch.

Dark matter

The thing that will decide how the universe will end is gravity. If there is enough matter in the universe, there will be enough to stop the universe expanding and even make it shrink. If there is not enough matter, the universe will expand forever.

There is not enough matter in all the stars and galaxies we can see to stop the universe from expanding. But astronomers believe that there is a lot of matter in the universe that they can't see. It is called dark matter. They can't see it, but they can detect ts gravity.

Astronomers have estimated how much matter there could be in the universe. But it does not seem to be enough to hold back the fleeing galaxies. So it seems as if the universe could go on expanding forever.

Earth is a rocky planet that circles the Sun.

We live in cities on Earth's land areas, or continents.

The universe in scale: These images give an idea of how we humans fit into the universe. The bottom line is that in the universe as a whole, we are not very important at all! And the planet we live on is but a grain of salt in a vast ocean of space.

The Mighty Universe

Above: One of the many clusters of galaxies found throughout the universe. It is a particularly interesting cluster, because it acts like a lens to bend light from a galaxy behind it. We see this galaxy as a series of blue arcs.

The Sun and billions of other stars belong to a galaxy, or great star island in space.

Billions of galaxies cluster together to make up the vast universe.

The Sun's family of planets and other bodies travel through space together.

43

EXPLORING STARS AND GALAXIES

Glossary

ACTIVE GALAXY A galaxy that gives out very much more energy than usual, often as radio waves or X-rays.

ATMOSPHERE The layer of gases around a heavenly body.

ATOMS The smallest bits of a substance. Every atom has a center, or nucleus, with electrons circling around it

BIG BANG A fantastic explosion that astronomers think created the Universe about 15 billion years ago.

BINARY A two-star system, in which two stars circle around each other, bound by gravity.

BLACK HOLE A region of space with enormous gravity; not even light can escape from it.

BLUESHIFT A movement, or shift, in the lines in the spectrum of a star or galaxy toward the blue end. It indicates that the object is traveling toward us.

CELESTIAL SPHERE An imaginary dark globe that appears to surround the Earth. The stars seem to be fixed to the inside of the sphere.

CEPHEID A variable star that changes in brightness regularly over a few hours or a few days.

CLUSTER A group of stars or galaxies. *See* open cluster, globular cluster, supercluster.

CONSTELLATION A group of bright stars that appear to form a pattern in the sky.

CORE The center part of a body.

COSMOS Another word for the universe.

DOUBLE STAR A star that looks like a single star but is actually two stars close together.

ECLIPSING BINARY A kind of variable star. It is a binary (two-star) system in which the stars regularly eclipse, or pass in front of, one another. This causes the brightness of the system to vary.

ELLIPTICALS Elliptical galaxies, which are round or oval in shape.

GALAXY A "star island" in space. Our own galaxy is called the Milky Way.

GLOBULAR CLUSTER A globe-shaped group containing hundreds of thousands of stars.

GRAVITY The pull, or force of attraction, that every body has because of its mass.

HEAVENS The night sky; the heavenly bodies are the objects we see in the night sky.

INTERSTELLAR Between the stars.

INTERSTELLAR MATTER Gas and dust found between the stars.

IRREGULARS Irregular galaxies, which have no definite shape.

LIGHT-YEAR A unit astronomers use for measuring distances in space. It is the distance light travels in a year—about 6 million million miles (10 million million kilometers).

MAGNITUDE A measure of a star's brightness.

MILKY WAY A faint band of light seen in the night sky. Our galaxy is also called the Milky Way.

Glossary

NEBULA A cloud of gas and dust in space

NEUTRONS Tiny particles found in the nuclei (centers) of most atoms.

NEUTRON STAR A very dense star made up of neutrons packed tightly together.

NOVA A star that brightens suddenly and appears to be a new star. Nova means "new."

NUCLEAR REACTION A process that involves the nuclei (centers) of atoms.

NUCLEAR FUSION A nuclear reaction in which light atoms (such as hydrogen) combine, or fuse, together. It releases enormous energy.

NUCLEUS The center of an atom.

OPEN CLUSTER A group of up to a few hundred stars that travel through space together.

ORBIT The path in space one body follows when it circles around another, such as a planet's orbit around a star.

PLANETARY NEBULA A cloud of gas and dust puffed out by a dying star.

PULSAR A rapidly spinning neutron star, which flashes pulses of light or other radiation toward us.

QUASAR A body that looks like a star but is much farther away than the stars and is as bright as hundreds of galaxies.

RADIATION Rays. The heavenly bodies give off energy as radiation—as light rays, infrared rays, gamma rays, X-rays, ultraviolet rays, microwaves, and radio waves.

RED GIANT A large red star. Stars swell up to become red giants when they begin to die.

REDSHIFT A movement, or shift, in the lines in the spectrum of a star or galaxy toward the red end. This indicates that the object is moving away from us.

SPECTRAL LINES Thin, dark lines in the spectrum of a star.

SPECTRUM A color band produced when the light from the Sun or a star is split into its separate wavelengths (colors), for example by passing it through a prism.

SPIRALS Spiral galaxies; galaxies with a spiral shape.

STAR A huge ball of very hot gas, which gives off energy as light, heat, and other radiation.

STELLAR To do with the stars.

SUPERCLUSTER A large grouping of clusters of galaxies.

SUPERGIANT The biggest kind of star, typically hundreds of times bigger across than the Sun.

SUPERNOVA The explosion of a supergiant star.

UNIVERSE Space and everything that is in it—galaxies, stars, planets, moons, and energy.

VARIABLE A star whose brightness varies.

WHITE DWARF A small, dense star; stars like the Sun eventually turn into a white dwarf when they die.

Exploring Stars and Galaxies

Important Dates

150 BC About this time the Greek astronomer Hipparchus catalogs more than 1,000 stars and introduces the scale of magnitude for estimating star brightness.

AD 150 About this time Ptolemy writes an encyclopedia of the scientific and astronomical knowledge of the day, containing the idea of an Earth-centered universe.

1054 Chinese astronomers record a supernova in the constellations Taurus (Bull), which gave rise to the Crab Nebula we see today.

1543 Polish churchman/astronomer Nicolaus Copernicus publishes his idea of a solar system, overturning the accepted Earth-centered view of the universe.

1609 Italian astronomer Galileo first observes the heavens in a telescope.

1668 Isaac Newton in Britain builds a reflecting telescope.

1675 Greenwich Observatory founded in England.

1781 French astronomer Charles Messier publishes his list of star clusters and nebulas; many clusters and nebulas (and galaxies) are known by their Messier (M) numbers.

1802 English astronomer William Herschel discovers binary stars.

1888 US astronomer J.L. Dreyer publishes New General Catalog of star clusters and nebulas; clusters, nebulas and galaxies are usually known by their NGC numbers.

1912 US astronomer Henrietta Leavitt studies Cepheid variables and establishes a relationship between their true brightness and period (time in which they change brightness).

1917 Completion of the giant 100-inch Hooker reflector at Mount Wilson Observatory, near Los Angeles.

1923 US astronomer Edwin Hubble, observing with the Hooker, proves that galaxies are distant independent star systems.

1931 US engineer Karl Jansky detects radio waves coming from the heavens and launches the science of radio astronomy.

1957 Russia's Sputnik 1 launches the Space Age.

1967 Radio astronomers at Cambridge, England, discover pulsars.

1990 Hubble Space Telescope launched.

1999 Sightings of planets around other stars.

2000 NEAR-Shoemaker goes into orbit around asteroid Eros

For More Information

Further Reading

Large numbers of books on astronomy and space are available in school and public libraries. Librarians will be happy to help you find them. In addition, publishers display their books on the Internet, and you can key into their websites and search for astronomy books. Alternatively, you can look at the websites of on-line bookshops (such as Amazon.com) and search for books on astronomy and space. Here are just a selection of recently published books for further reading.

Constellations by Paul P. Sipiera, Children's Press, 1997
Eyewitness: Astronomy by Kristen Lippincott, Dorling Kindersley, 2000
Map of the Universe, Smithsonian Institution, 1996
Monthly Star Guide by Ian Ridpath and Wil Tirion, Cambridge, 1999
Night Sky by Gary Mechler, National Audubon Society, 1999
See the Stars by Ken Croswell, Boyds Mills Press, 2000
The Sky at Night by Robin Kerrod, New Burlington Books, 2000
Stargazing by Patrick Moore, Cambridge, 2000
Turn Left at Orion by Guy Consolmagno and Dan M. Davis, Cambridge, 2000
The Young Astronomer's Activity Kit, Dorling Kindersley, 2000

Websites

Astronomy and space are popular topics on the Internet, and there are hundreds of interesting websites—details about the latest eclipse, mission to Mars and SETI (Search for Extraterrestrial Intelligence), and so forth.

A good place to start is by using a Search Engine, and search for space and astronomy. Search engines will display extensive listings of topics, which you can then select. For example, you gain access to a list of topics on the Search Engine Yahoo on astronomy with: **http://yahoo.com/Science/Astronomy**

The lists also includes astronomy clubs. If there is one near you, you may well like to join it. Most clubs have interesting programs, with observing evenings, lectures, and visits to observatories.

NASA has many websites covering all aspects of space science, including exploration of the planets and the universe as a whole. The best place to start is at NASA's home page: **http://www.nasa.gov**

From there you can go to, for example, Space Science, which includes planetary exploration. Or you can go directly to: **http://spacescience.nasa.gov/missions**

Individual missions may also have their own website, such as the Mars Odyssey mission at: **http:/mars.jpl.nasa.gov/Odyssey**

The latest information and images from the Hubble Space Telescope can be reached at: **http://www.stsci.edu/pubinfo** This site will also direct you to picture highlights since the launch of the Telescope in 1990.

European space science activities can be explored via the home page of the European Space Agency at: **http:/www.esa.int**

EXPLORING STARS AND GALAXIES

Index

Alcor; 14
Aldebaran; 16
Algol; 9
Alpha Centauri; 14
Andromeda Galaxy; 30, 35-36

Betelgeuse; 9
Big Bang; 5, 40-41, 44
Big Dipper; 14
black hole; 37, 44
blueshift 13; 40, 44
Brown dwarf; 21

Centaurus; 8, 16
Cepheids; 9, 30
cluster; 5, 15-16, 30, 32-34, 36, 38, 42, 44-45
Copernicus; 38
Crab Nebula; 25
Cygnus X-1; 29

Dog Star; 23
double star; 14, 44

helium; 11, 22
Hercules; 16
Hipparchus; 8
Horsehead Nebula; 17
Hubble, Edwin; 30, 32
Hubble Space Telescope; 26
Hyades; 16
hydrogen bomb; 11

Large Magellanic Cloud; 26, 35
Leavitt, Henrietta; 9
light-year; 6, 8, 17, 30, 34-37, 41, 44

M13; 16
Magellanic Clouds; 36
Milky Way; 34-36, 44
Mizar; 14-15
Mount Wilson Observatory; 30

nebula; 5, 17, 23, 26, 30, 45
neutron; 27-28, 45
neutron star; 27-28, 45
Northern Hemisphere; 15-16, 35
nuclear fusion; 11, 45

Omega Centauri; 16
Orion; 9

Perseus; 9
Pleiades; 15-16
Proxima Centauri; 8

rainbow; 13
red dwarf; 21
redshift; 13, 40
Ring Nebula; 23

Seven Sisters; 15
Sirius; 8, 14, 23
Sirius; 6
spectrum; 13, 40, 44
supercluster; 42, 44, 45
Supernova 1987A; 26

Taurus; 16, 25
Tucana; 16

Ursa Major; 14

white light; 12